MW00718594

INNOVATORS
Rock Stars of STEM
Science, Technology, Engineering, and Mathematics

by Martha J. LaGuardia-Kotite

Eglin Air Force Base · Florida · USA

Design by Joe Fahey
Edited by Eric Braun

Front cover photos by (top left) AFRL/RW, (top right) Greg Vojtko/DMDS, (bottom left) Kenneth Stewart/DVIDS, (bottom right) US Government. Back cover photo by AFRL/RW.

Air Force Armament Museum Foundation
100 Museum Drive
Eglin Air Force Base, FL 32542
www.afarmamentmuseum.com
(850) 830-3321
Or contact Martha J. LaGuardia-Kotite directly at:
www.marthakotite.com

Printed and bound in the United States of America

First Edition

10 9 8 7 6 5 4 3 2 1

LCCN 2015947978

ISBN 978-0-692-49797-5

This book was expertly produced by Book Bridge Press.
www.bookbridgepress.com

Contents

Innovation and Collaboration

"The people who are crazy enough to think they can change the world are the ones who do."

—Apple's "Think Different" commercial, 1997

Photo by Jon Snyder/Wired.com

Steve Jobs cofounder, chairman, and CEO of Apple Inc.

Who are the most successful innovators of our time?

Many would put Steve Jobs near the top of the list. When he was only 21, he co-founded Apple Computer with his high school classmate in his parents' garage. He pioneered groundbreaking computer technologies with the Macintosh, iPod, iPhone, and iPad. His powerful music app, iTunes, changed the music industry. His work also transformed animated movies, digital publishing, and retail (the way we buy and sell things).

By making technology and computers smaller, cheaper, and more user-friendly, Jobs has been credited with revolutionizing the computer industry. Though he died in 2011, we continue to feel the effects of his legacy. He's an example of the amazing things women and men can do when they use their imagination and creativity while mixing art with technology. *Innovators think differently.*

Like Steve Jobs, the innovators in this book combine their creativity and determination with skills in one or more of the areas of science, technology, engineering, math, and medicine (STEMM). They have made important impacts on the world by finding ways to keep soldiers safer, fight and win wars, and create new technology that will change how the military and businesses work.

Photo by Greg Vojtko/DVIDS

Naval Surface Warfare Center Corona Division engineer Erika Garcia, right, directs students in the use of ground tracking system during the 12th annual Science Technology Partnership conference.

All of the scientists in this book found rewarding careers at the Air Force Research Laboratory at Eglin Air Force Base in Florida, but workers in the STEMM fields are needed in many industries all over the world. Bright, determined people can find jobs in biology and nature research, cyber or robotics, animated films, astronomy and space exploration, climate and marine research, and many other careers that improve human lives.

This book is about people who were driven to find a better way. Each one of these scientists took a spark—an idea—and shaped it into reality. And sometimes, they even changed the way we live and think.

Image public domain

Chapter 1

Dream It! Make It! Fly It!

Aeronautical engineer Ken Blackburn

Photo from author's collection

Ken Blackburn is an aeronautical engineer. He used his knowledge of paper airplanes to design a groundbreaking, lightweight, low-speed aircraft or drone called BATCAM.

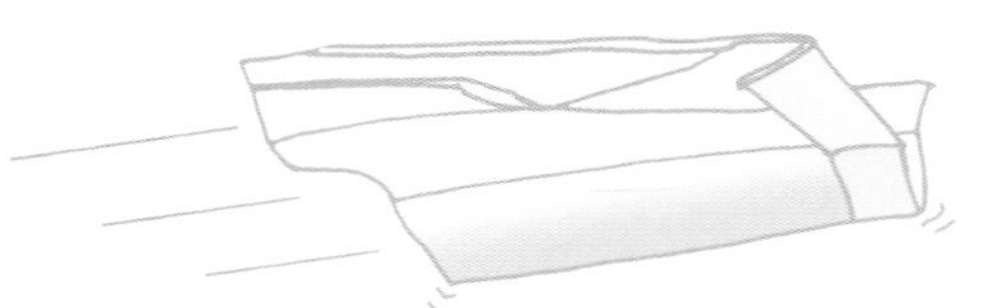

At the age of 13, Ken Blackburn designed a new kind of paper airplane. His plane was able to fly so high and for so long, circling slowly to the ground, that it would later earn him a place in the book of *Guinness World Records*—four times.

Blackburn first learned how to make paper airplanes from friends in elementary school. He quickly became fascinated with airplanes and flight. He went to the library and checked out every book he could find on paper planes. He read about real airplanes, too. He made lots of paper airplanes and tested them in flight.

"I was always known as the paper airplane guy," Blackburn says. "I tried to learn what I could about real airplanes and apply it

"I was always known as the paper airplane guy."

to paper airplanes because it's the same physics. If you can understand what makes a paper airplane fly, you can understand what makes an F-35 [fighter jet] fly."

Blackburn is an aeronautical engineer and team leader at Eglin Air Force Base in northwest Florida. There, he has used his knowledge of paper airplanes and the principles of flight to develop a groundbreaking unmanned aerial vehicle (UAV) called BATCAM. A lightweight, camera-equipped, low-speed aircraft—or drone—BATCAM is about the size of a large raptor such as an osprey. Soldiers use it to scout dangerous areas without putting themselves at risk.

"The BATCAM is a flying robot," Blackburn says. "There are some unique challenges with low-speed flight. Thin wings are better at low speed than real thick wings, which is one reason why butterflies and insects have thinner wings than birds. The BATCAM has very thin wings, and so do paper airplanes. It flexes like a bird," he says, explaining how the BATCAM withstands powerful wind gusts. The

Photo by Jim Hazeltine/US Air Force

56th Operations Group flagship F-16 Fighting Falcon escorts Luke Air Force Base's first F-35 Lightning II to the base.

Photo courtesy of US Air Force

Battlefield Airman Targeting Camera Autonomous Micro-Air Vehicle (BATCAM) during a test flight.

Photo courtesy of US Air Force

Colorized Scanning Electron Microscope image of an insect wing.

BATCAM's wings are so thin, in fact, soldiers can roll it up and slip it into a tube.

Blackburn is an expert in low-speed aerodynamics and weapon systems design. He has worked to find ways to make small UAVs better. He describes in one of his research papers how wing flexibility can improve the small drone's camera steadiness. In another, he investigated using a rotating birdlike tail for control of the UAV. His work also included analysis of how to improve the range and speed of a jet-powered UAV.

Blackburn grew up on a farm in Winston Salem, North Carolina. "I was very quiet. I was always interested in science," he says.

Occasionally his family would go to the beach. There he saw the memorial to the Wright Brothers on the grounds where they made their first flight. This furthered his interest in aeronautics. To him, flight seemed magical. Who made airplanes? What made them work? Where did airplanes come from? After he designed his own paper airplane, he dreamed of becoming an aeronautical engineer.

When Blackburn was in college at North Carolina State University, his dorm mates encouraged him to use his paper airplane design to try to set a world

Blackburn has designed about 200 paper airplanes and made over 10,000 out of "almost any kind of paper product—from Post-it Notes to pizza boxes." But he makes most of his planes out of regular printer paper.

Photo from author's collection

Ken Blackburn designed a new kind of paper airplane that flew so high and so long before landing that it earned him a place in the book of Guinness World Records *four times. This is an example of that record-making paper airplane.*

record. He decided to give it a shot. With a flying time of 16.89 seconds, he did set the record, which was recorded in the 1983 book of *Guinness World Records*.

He graduated from college in 1985 with a degree in aeronautical engineering. He worked for McDonnell Douglas in St. Louis, Missouri, as a researcher for fighter aircraft fueling systems. In 1987, he set another paper airplane record with a flight lasting 17.2 seconds.

Blackburn continued to improve and test his original design. "Four or five days a week I would practice making paper airplanes and making adjustments. I made ridges like the dimples on a golf ball to reduce drag. [The ridges] stir up the air to allow the air to stick to the wing better."

In 1994, he again set the world record with an 18.8-second journey. Finally, on October 8, 1998, he sent a paper airplane skyward inside the Georgia Dome in Atlanta for a world record of 27.6 seconds aloft.

In spite of his world records, Blackburn has failed along the way, too.

"I can think of five or six times where I tried to set a record and did not," he says. "I learned from each attempt, but the biggest lesson is persistence—it pays off. If you are willing to work longer, harder, and smarter, you can still do better. You can be as good as you want

How Do Planes Fly?

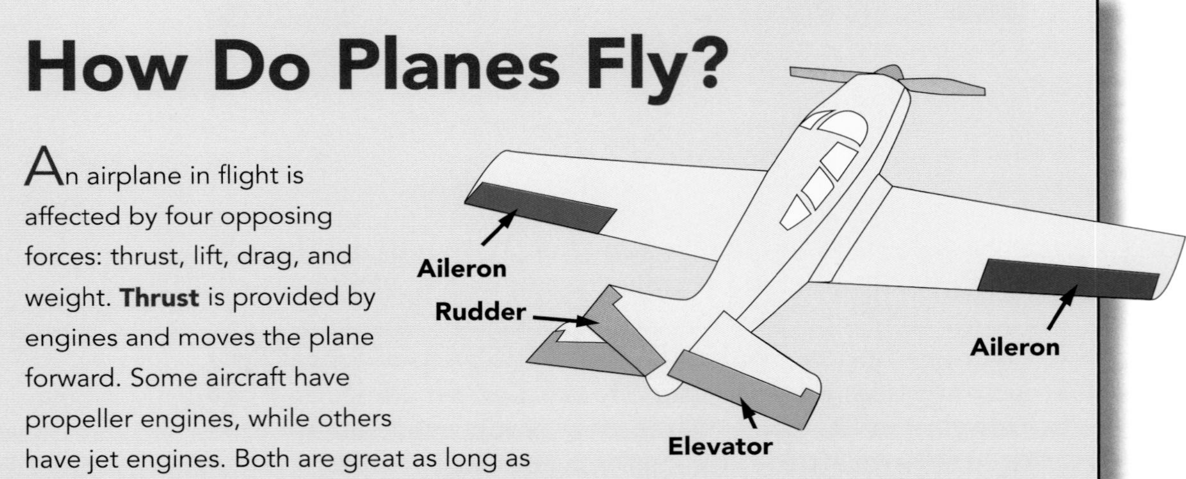

An airplane in flight is affected by four opposing forces: thrust, lift, drag, and weight. **Thrust** is provided by engines and moves the plane forward. Some aircraft have propeller engines, while others have jet engines. Both are great as long as they work hard enough to make air flow rapidly over the wings, splitting the air.

Air moving under the wing toward the ground creates a force called lift. **Lift** pushes up, giving the plane's wings a boost, lifting the plane skyward. **Drag** is the resistance of the air against the aircraft. It works against thrust and slows the airplane. On a windy day, walking into the wind feels like drag. The shape of the wings is designed to reduce drag and let air flow around them smoothly. The last force is the weight of the airplane. **Weight** is affected by gravity and pulls the plane down toward the ground.

In order to fly, an aircraft has to control all four forces very carefully. It does that with its engines, wings, ailerons, rudder, and elevators.

Even with all those factors, one more element is needed: a pilot. How about you?

In 1988 Ken Blackburn set the paper airplane world record for 27.6 seconds aloft.

to be. It's all a matter of how much you're willing to practice and work at something."

The BATCAM was a result of Blackburn's hard work and a lifetime of experimenting with small, low-speed paper airplanes and designs. As a kid, though, he only imagined designing planes with people in them. The growing importance of UAVs opens up a whole new career field that excites Blackburn.

"You can do things with unmanned vehicles that you can't do with manned vehicles," he says. "UAVs can fly longer than a human can stay awake. And because they're small, they can go where manned aircraft cannot fit."

Beyond the military purposes, UAVs are used for aerial surveillance of crops, filmmaking, and assisting with search-and-rescue operations. They're also used to count and protect wildlife from poachers, deliver supplies to remote areas, monitor crowds, and even take real estate photographs of homes for sale.

Though Blackburn's BATCAM is a groundbreaking innovation, he is still best known for his paper airplanes. That's true even though a Japanese engineer broke his record in 2009 with a 27.9-second flight.

During his lunch break at the air force base, Blackburn finds time to enjoy his hobby inside an unoccupied hangar. Here, he practices throwing paper airplanes and considers reclaiming his world record.

"I'm a four-time world record holder," Blackburn says. "I have no complaints. If I can get the record back, it's a bonus." Though he isn't the current record holder, he is a sought-after paper airplane judge for tournaments around the United States and Austria.

"Paper airplanes opened up all kinds of opportunities I never would have had otherwise," he says.

Chapter 2

Secret World Scientist

Research scientist Nydeia Bolden-Frazier, Ph.D.

Photo from author's collection

Dr. Nydeia Bolden-Frazier is one of the nation's leading research scientists helping to design military equipment. She works in a secret world perfecting weapons systems.

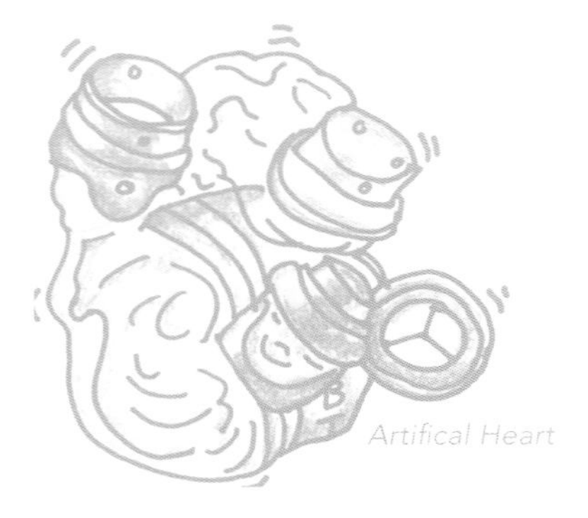

Nydeia Bolden-Frazier works in a secret, enclosed world, perfecting deadly weapon systems, surrounded by people she affectionately describes as "super-smart geeks." But Bolden-Frazier doesn't consider herself to be "super smart."

"I have interests in certain areas," Bolden-Frazier says, explaining her success. "I'm dedicated to whatever choice I make, and I work hard at it."

That's a modest statement for one of the nation's leading scientists helping design the military's top guns, bombs, and equipment.

"It's not a certain type of person. It's a matter of drive. If you have the drive and are interested, you can do it."

"As a scientist, I study how and why things happen. Here, I use science to make the best weapons in the world."

Bolden-Frazier is the Assistant Chief Scientist for the Air Force Research Lab at Eglin Air Force Base. She provides guidance to more than 200 scientists and engineers. Her work aims to identify problems and create scientific solutions, which are tested in the lab. One project she's working on is a self-healing material that automatically repairs weapons before small damage becomes too big. This solution is based on how human skin self-heals.

"The problem we're trying to solve occurs when there are very small cracks on a weapon which cannot be seen until the material breaks apart. Then, it's too late," she says. "The self-healing process kind of works like the body. The body knows when it's damaged and heals itself."

The eldest child of a brick mason and a dietician, Bolden-Frazier grew up in Daphne, Alabama, with four sisters and a brother. Her birth parents divorced when she was 13. Bolden-Frazier's down-to-earth style developed from tackling challenges in life and in school.

An assignment in high school changed Bolden-Frazier's life. She wrote a report about an artificial heart a team of doctors and biomedical engineers developed.

"I was just so fascinated they developed this artificial heart and put it into a body, and the body continued to work," Bolden-Frazier says. "It was a defining moment. I realized I wanted to go into biomedical engineering."

She attended the University of Alabama-Birmingham campus, planning to focus on biomedical engineering. Instead, she received her degree in materials science and engineering, specializing in metallurgy (the study of metals). She went on to earn a master's degree in chemistry and a doctorate in materials science from Tuskegee University. These degrees make her highly respected in scientific and academic places. Yet, even having earned the right to be called "Doctor Bolden-Frazier," she says, "Please, call me Nydeia."

A year and a half before graduating in 2009, she accepted the position as a materials engineer for the lab at Eglin Air Force Base, where she started developing the self-healing material.

Bolden-Frazier and a team of scientists asked questions to create a solution: *How does a material know it is damaged? When does the material know it needs to fix itself?*

The product they're developing is like scratch-resistant paint used on automobiles. "When a car's paint is damaged, micro-capsules of paint break open and release paint, filling the holes," Bolden-Frazier says. "It looks like it was never damaged. The idea is to enhance the reliability of weapons. You want them to function as intended on demand."

Bolden-Frazier took this concept into the lab, applying the solution to small drones, or unmanned

Air Force Research Lab

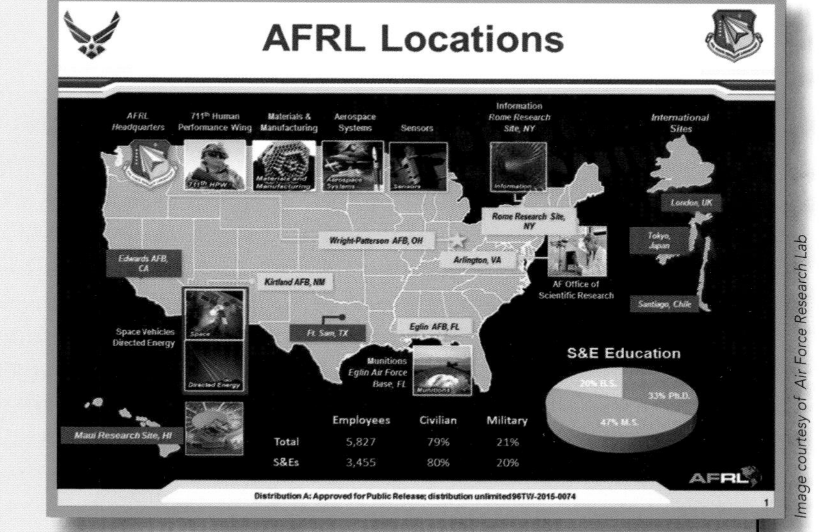

Image courtesy of Air Force Research Lab

Many Air Force Research Labs are located around the United States. There's aerospace and sensors research at Wright-Patterson AFB in Ohio. There's scientific research in Arlington, Virginia. The munitions lab is located at Eglin Air Force Base where Dr. Bolden-Frazier works.

Just inside the entrance to her building there's a room, like a museum. There you can see the munitions, weapons, and devices the lab has developed. On one side there's the BATCAM, which can be rolled up and put in a tube. This lab developed software to launch the small drone with a GPS video link. In another corner, there's a collection of fuzes and pieces of shrapnel, like shredded metal—remains from the explosion of a bomb. The fuze is designed to kick off the explosion of the bomb it's attached to at a precise time, speed, and depth inside the object, like a concrete building or bunker. These scientists are hoping for their experiment to give the most destructive result.

Chapter 2

Photo from author's collection

Dr. Bolden-Frazier with her husband and son.

aerial vehicles (UAVs). “If a soldier drops the backpack that's holding the UAV, it doesn't need to be inspected for cracks,” she says. “In 20 minutes, the device's material heals itself— ready to go.” The material is still in development and an ongoing project.

When Bolden-Frazier is not at work, she likes to explore the outdoors. She hunts deer, turkey, and small game such as squirrel, taking her six-year-old son with her. He has two .22-caliber rifles and a Jr. Longbow for hunting. “He loves it,” Bolden-Frazier says.

Picking her greatest achievement so far, she says, is tough. Finally she settles on earning her Ph.D. “Because it never was my plan. I had a stereotypical picture of a Ph.D. in my mind: pocket protectors, the geek. The really, really smart person.”

One thing she has learned along the way is how to approach her career goals one step at a time.

“I need to do A to get to B, so I focus on A,” she says.

To finish her undergraduate degree she had to overcome hurdles. During her senior design project, she had to choose materials to redesign a product.

“At the last minute the deadline changed, and I was in a time crunch. A guy who was in the Ph.D. program was helping me figure out what I needed to do to finish the project. I didn't do the background work I should have. I took his word for it. I did the presentation and it was all wrong. In the first five minutes I could see my professor's facial expression, *no, no*. This was like an epic fail."

In front of the class, the professor asked her where she got her information. She told him, and he said, "One of the first things I've told you all is you have to have a reliable source."

Later, she contacted her professor and admitted her mistake. The professor agreed to let her present again. She earned a "B" and graduated.

If Bolden-Frazier hadn't gone into science, she says she would have become a mechanic.

"I like to see girls doing typical male jobs," she says. "I love to see female fighter pilots."

One of her next projects is Camp NERDS (Next-gen Engineers, Researchers, and Developing Scientists). Working with the Doolittle Institute Innovation Lab, she'll host a summer camp for elementary school students. The weeklong program will introduce children to engineering and science ideas.

One of her hopes when she meets new people is that they think not of her qualifications, but think of her as just like anybody else.

"I like to inspire kids. Make a kid feel like he or she can do it," she says. "It's not a certain type of person. It's a matter of drive. If you have the drive and are interested, you can do it."

Chapter 3

Big Guns, Bombs, and Robots

Engineering technician Charles "Mac" McClenahan

Charles "Mac" McClenahan stands with one of his many successful projects after receiving the Air Force Meritorious Civilian Service Award.

Charles "Mac" McClenahan knows about guns—big guns! He knows about bombs, fuzes, and robots, too. The guns he has developed and cares for fire from aircraft to ground targets as well as from the ground into the sky. His bombs drop from planes with fuzes he's helped develop to blow up those bombs. He even created robots to recover unexploded bombs in the field.

"We've got the best job in the air force," Mac says of his dream job, working as an engineering technician for the Fuzes Branch of the Air Force Research Lab. He takes an idea for making weapons that go deeper into a target

"Try to learn something new everyday."

and explode faster, and he builds the device. All the while he's helping the US Air Force fight battles more efficiently.

Mac's office is tucked inside a bomb laboratory of sorts—a single-story cinderblock building nestled on the edge of Eglin Air Force Base. He has a warehouse filled with casings for various sizes of missiles and mechanical equipment to cut, shape, and create. Just out back, in a huge field of orange clay and wild grass, he's got chunks of cement targets—some whole, some cracked and holed like leftovers from a military video game.

Nearby, three M110 tanks rest in the sunlight. Mac walks across the sand-and-gravel driveway and through a field scattered with wild raspberry bushes toward

Photo from author's collection

Near Mac's bomb laboratory is the field of orange clay and chunks of cement targets that he and his team fire weapons into to test their power and speed.

Chapter 3

Photo from author's collection

Inside a steel hangar, Mac leans on one of the many big guns that he helped alter to perfect the weapon's ability to destroy targets. This is STUBBI, a 155-mm Howitzer built in 1955.

an arch-shape steel hangar. He carefully opens the metal door. The sunlight penetrates the deep, dark hold. Brightened by the rays of light, big guns rest on the cool gravel like props from a war movie. Smiling, Mac's gaze rests on the collection.

These short-barrel cannons designed to fire shells high into the air at short distances have colorful names: "Stubby House," a 155-mm Howitzer, was built in 1955; "Dragon's Breath" is a slightly smaller Howitzer, at 105-mm. Mac uses them to test fuzes.

His pride and joy is a gun truck he helped design, equipped with a 155-mm Howitzer. It can be fired remotely. It's used for testing "runway penetrators," or missiles that explode runways. "It will recoil about 43 inches and raise the truck up about 12 inches off the ground," Mac says with a wide grin. Another gun he helped create, a 105-mm mounted inside an AC-130 aircraft, will jolt the tail of the aircraft about two feet. "Kick it really hard," Mac says.

With only a high school diploma, Mac became a renowned armament-systems expert. He shaped history for the men and women who serve in combat by coming up with ideas for improving weapons, investigating concepts, and creating solutions. He also modified several 1950s-era gun systems so they could use modern ammunition, which is more widely available and less expensive. This saved the military millions of dollars.

Mac was born in Pennsylvania in 1946, a year after World War II. His parents, originally farmers, moved into Grove City, Pennsylvania, a town of 8,000 people. "They had ten cents in their pockets," Mac says. His dad took a job as a foreman building diesel engines for Cooper-Bessemer Industry.

As a child, Mac remembers getting measles. His house was quarantined. To pass the time, he played with Lincoln Logs and Erector Sets, creating buildings, cranes, and cars. His father also worked part-time wiring houses for electricity. From the time he was eight years old, Mac crawled into small areas to run the wires for his dad.

His father showed him how to build things, too. Together they built offices, a boat, and a metal roof for a bus barn. Mac made gym equipment for the high school. After high school, he joined the US Air Force.

His life in the air force included two years as a munitions specialist and 20 years as an Explosive Ordnance Disposal (EOD) specialist. He was assigned to locate unexploded bombs and, using his weapons training and knowledge, defuse them. He served in Thailand and at some of the bombed sites in Laos during the Vietnam War. He helped clear the unexploded bombs to resettle a village. He was also assigned to work in Turkey constructing a bombing range.

Mac's warehouse is filled with casings for various sizes of missiles and mechanical equipment to cut, shape, and create bombs for the military.

Photo from author's collection

Chapter 3

Mac spent most of his career at Eglin Air Force Base in Florida where he learned what he needed to know about electronics and all kinds of weapons. His experiences helped fine-tune his ability to recognize a problem, come up with an idea, and develop a workable solution.

Mac says he wanted to do more than defuse unexploded bombs. He wanted to figure out why they didn't explode in the first place. "So I came up with robots," he says.

Mac's Robots

Mac's robots have been useful in other areas, too. When an accident at the Three Mile Island nuclear power plant damaged the reactor in 1979, he was called to the scene near Harrisburg, Pennsylvania. He used a remote vehicle to "flip some switches" to help stabilize the scene. Even NASA has called upon his expertise. Mac helped test a spacecraft design for a future mission to Mars.

"We built a specialized cannon for the gun truck that launched the experimental Mars Probe into simulated Martian soil to test survivability of the probe," Mac says. "The launcher tested the capsule that would go into Mars. This was the very first step to putting one up there."

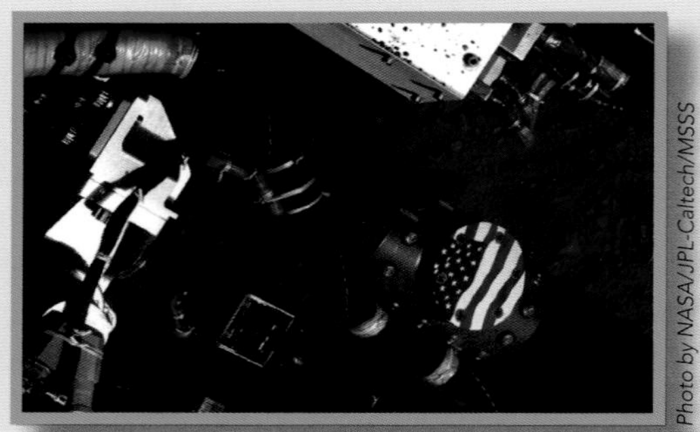

Photo by NASA/JPL-Caltech/MSSS

(Top) American flag medallion on NASA's Mars rover Curiosity. *(Bottom) Back shell of NASA's InSight spacecraft being lowered onto the mission's lander.*

Photo by NASA/JPL-Caltech/Lockheed Martin

He created the EOD Remote Ordnance Recovery System that located, dug up, and removed the fuzes from unexploded bombs. These robots could remove the fuzes from bombs that weighed up to 2,000 pounds and bring them back for analysis: Why did the device malfunction? Mac's robots have helped the military build more reliable bombs.

Of all his accomplishments, his proudest one seems simple. AC-130 aircraft typically had a 40-mm gun mounted on them. But the military was running out of 40-mm bullets and parts to maintain the guns. Mac came up with the solution: Take the 40-mm gun out and convert it to the 30-mm gun, which was widely available. His challenge was persuading his bosses to give him a chance to convert the old into something new. He persisted. Six months later, he demonstrated the result—a gun that's affordable to shoot, that has widely available parts, and that's accurate.

"The 30-mm gun is what's going on all the gun ships from now on," Mac says.

Mac believes his success is a result of collaboration. "If you have an idea, you think about it as far as you can and then start surrounding yourself with smart people to get you over the hill," he says. "The 30-mm gun mount is an example. I could not do the stress analysis on it. I had to go to an engineer to do that."

Mac understands the importance of an education in science and technology fields. "There's people like me, we've got the ideas and we don't have the education. It takes a combination of those two in order to come up with a product or a program."

Even though he doesn't have a degree, Mac has worked hard to learn. As he says, "Try to learn something new every day."

Chapter 4

It's About Time

Munitions technologist Mikel Miller, MS, Ph.D.

Photo from author's collection

An expert of time and navigation, Dr. Mikel Miller has studied nearly every method of navigation. He's especially fond of GPS, a satellite navigation system developed by the United States. It evolved from military purposes, like guiding aircraft and missiles, to everyday commercial uses found in cars, cell phones, and cameras.

"Time is one of our most precious resources," Mikel Miller says. "Scientists have always sought to protect and maximize it."

Miller is an expert on time and navigation—figuring out exactly where you are and how to get where you want to be. He has studied nearly every method of navigation in history, from ancient methods to today's Global Positioning System (GPS). With this expertise, he directs all the research and development being done in the lab at the Munitions Directorate at Eglin Air Force Base. He is the main scientist responsible for meeting all the air force's time and navigation needs.

"Time is one of our most precious resources. Scientists have always sought to protect and maximize it."

Navigating by Satellite

GPS is the satellite navigation system developed by the United States. Many other countries have their own versions. At first, these systems were used only for military purposes—guiding aircraft and missiles. But not anymore.

Photo courtesy of United States Government

GPS Block IIF satellite

"Now you can buy cameras with GPS that provide better than five meters of accuracy," Miller says. GPS has not only gotten more accurate, it's also gotten easier to use. Once the size of an ice-cream truck, today's receivers can be as small as a thumbnail. "We now have miniature GPS receivers in our cars, cell phones, and cameras!"

Photo by Lance Cpl. David Flynn/DVIDS

Atomic Clock

Photos courtesy of US Naval Observatory

In Washington, DC, the US Naval Observatory serves as the official source of time for the US Department of Defense and the standard of time for the entire United States. The observatory's master clock consists of several dozen individual atomic frequency standards that form a "backbone" time-scale, steering the output of a single hydrogen maser. These two images show the master clock's "top end" and a typical clock vault containing some of the many clocks that are used in the ensemble to keep track of time.

So what does this have to do with time?

"The most important use of GPS is precision time," Miller explains. "It's the cheapest method for determining time to the nanosecond level. All commerce, banking transactions, cell phone towers, power grids have to be linked when it comes to precision communications for timing."

GPS uses atomic clocks to keep time. These keep time better than any other clock. Without an atomic clock, GPS navigation would be impossible. The United States Naval Observatory (USNO) has the mission to supply the very precise time used by GPS. This master clock in Washington, DC, synchronizes all the atomic clocks placed in each GPS satellite. GPS satellites distribute this precise time to users.

"If time gets messed up, everything gets messed up," Miller says. More errors would be introduced into navigation.

In high school, Miller loved math, science, and sports, especially pickup baseball games. The son of an enlisted air force veteran and a German mother, he and his older sister and brother moved around a lot. While living in North Dakota, during his junior and senior years of high school, he was inspired to seek a career in engineering thanks to his math and chemistry teachers. They made the subjects exciting and challenging.

Miller attended North Dakota State University in Fargo, earning a Bachelor of Science degree in electrical engineering in 1982. He joined the US Air Force, first working as a satellite systems engineer in Nebraska. Later he worked in research, development, and program management. It was during the 1980s when he got to be part of a big change in navigation history.

Photo by PA1 Donnie Brzuka/USCG

USCG Academy cadets learning to navigate with sextants, which can be used to plot a ship's postion using stars or the sun.

From INS to GPS

Before GPS was developed, spacecraft, missiles, surface ships, submarines, and aircraft used Inertial Navigation Systems (INS). INS uses different kinds of motion sensors within the vehicle to navigate. The instruments detect every move the vehicle makes and feeds it to a computer program. Based on where the vehicle started out, the computer uses the input from the instruments to understand where it is and where it should go. But INS

Photo by Charles Gaddis IV/DVIDS

Navy cadet plotting GPS with compass.

is not perfect—it's easy to introduce errors.

The consequences of those errors can be deadly. Over 30 years ago, the Soviet Union shot down Korean Air Lines Flight 007, killing 269 passengers and the crew. Leading up to this 1983 tragedy, the Soviet Union had a tense relationship with the United States and its allies. Both struggled for supremacy in a period known as the Cold War. When the Korean airliner strayed over Soviet territory, the Soviets suspected the purpose was for spying.

"The pilots thought they knew where they were, but tragically, they had accidentally initialized their INS system to an incorrect latitude and longitude point," Miller says. "Thus, Flight 007's INS *thought* it took off from a different location. The plane drifted off its intended course and flew directly into Soviet Union airspace."

The incident sparked global outrage and encouraged President Ronald Reagan to announce a new space-based anti-ballistic missile system. The system was designed to prevent missile attacks from other countries, specifically the Soviet Union. The media called it the "Star Wars" system. Reagan also wanted to improve navigation so mistakes like that could be avoided in the future. That's where GPS comes in.

GPS is a satellite-based navigation system with 24 satellites placed in orbit by the US Air Force. The satellites revolve around Earth approximately twice a day. As they circle our planet in a precise orbit, they transmit signal information at the speed of light to Earth. GPS receivers use those signals to calculate the user's exact location by comparing the time a signal was transmitted with the time it was received. This difference tells the GPS receiver how far away the satellites are relative to the user. This information allows users to determine their position and see it displayed on an electronic map.

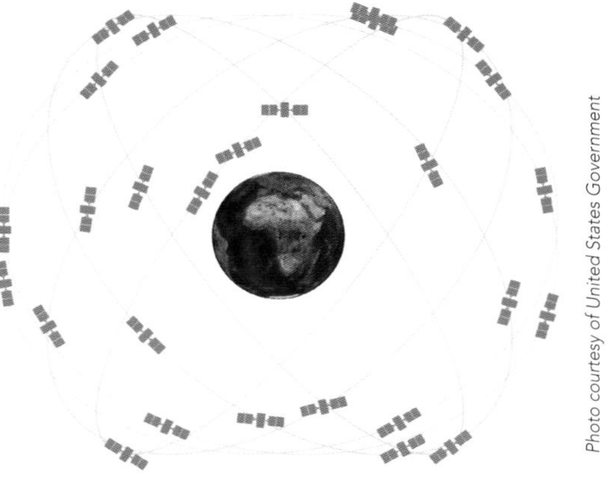

GPS constellation

Photo courtesy of United States Government

Can-Do Attitude

Miller earned a master's of science degree in electrical engineering in 1987 and a Ph.D. in electrical engineering in 1998. He retired from the air force as a lieutenant colonel in 2003 as one of the world's foremost experts on navigation. Since then, he has authored more than 60 journal articles and two handbooks on navigation technologies. He has been recognized with many prizes and honors, including being named a Fellow of the Institute of Navigation (ION) and the Royal Institute of Navigation (RIN).

When asked what it takes to be successful in a STEMM career, Miller says, "A positive, can-do attitude is critical. STEMM students and professionals are people who want to discover, innovate, and make the world a better place."

Miller has had that kind of attitude since he was a kid. Whenever he met a challenge, he'd find a hammer and a nail and build a solution.

Miller says, "If you have that desire to fix or build something, we need you in STEMM."

Afterword for Parents and Teachers

Paul Maryeski, Co-Director, Engineers for America

From the wheel to the World Wide Web, humankind's technical advancement throughout time has been truly remarkable. Today, technology turns over quickly with dramatic improvements in medicine, information management, aerospace, physics, and more.

The scientists profiled in *Innovators* have been on the front lines of these dramatic improvements. They stand out daily by showing exemplary commitment and resolve. Each, through his or her own experience, demonstrates "the art of the possible."

Engineers for America, now in its ninth year of operation, is a model program designed to motivate young students in both elementary and middle school toward technical careers. At the STEMM Center in Valparaiso, Florida, students conduct experiments, called "sorties," which are carefully designed. Scientific principles and processes are utilized. Data is recorded and results are analyzed. Then the students are provided a tour of exhibits at the Air Force Armament Museum. The exhibits are practical examples of engineering application and success. Volunteer engineers from Eglin Air Force Base conduct the student tours and answer many questions coming from the young and inquisitive minds. Parents are very much involved in the program, serving as aides during the experiment phase and as chaperones

Photo by Paul Maryeski

at the museum during the motivational tour. They help make it a fun and rewarding experience for the young "engineers." That is very instrumental.

Okaloosa County at large is stepping forward with key STEMM initiatives. The STEMM Center itself is testament to the vision and commitment of dedicated county leaders in both education and private industry. Northwest Florida State College is holding regular STEMM events for students in both middle school and high school. Professional organizations such as the American Institute of Aerospace and Aeronautics are very involved with STEMM outreach programs for all grade levels. There

Photo by Paul Maryeski

Air Force Armament Museum, Valparaiso, Florida

is a wealth of STEMM talent in the area and the collective energy and focus on results is making a difference.

Special thanks to the Air Force Armament Museum Foundation, the Okaloosa County School District, the Air Force Research Laboratory at Eglin Air Force Base, and the Doolittle Institute for making Engineers for America the dynamic and model STEMM program it is today. Since 2007, more than 5,500 students have participated in the program. Teachers are playing a most critical role. Their demonstrated interest in their students' advancement is most noteworthy. They are indeed the engines that make the program work.

It is most rewarding and for me a distinct privilege to be part of the day-to-day process toward building a smarter and stronger nation better able to meet tomorrow's economic and technological challenges.

Photo from 2012 Colorado All Service Academy Ball

About the Author

Martha J. LaGuardia-Kotite is a journalist who has written numerous books including the award-winning *So Others May Live: Coast Guard Rescue Swimmers Saving Lives, Defying Death*. Her stories have appeared in national and regional publications including *The Boston Globe, Yachting, Southern Boating* and *Emerald Coast Magazine*. A graduate of the US Coast Guard Academy and a captain in the reserves, she holds a master's degree in journalism from Harvard. She enjoys speaking with students, book clubs, and professional groups about her books and stories. *Innovators* is her fifth book. Connect with Martha at www.marthakotite.com, FB, and @MKotite.